DE HAVILLAND CANADA DHC-6, DHC-7, DHC-8

Paul R Smith

Copyright © Jane's Publishing Company Limited 1987

First published in the United Kingdom in 1987 by
Jane's Publishing Company Limited
238 City Road, London EC1V 2PU
in conjunction with DPR Marketing and Sales
37 Heath Road, Twickenham, Middlesex TW1 4AW, England

ISBN 0 7106 0473 4

Printed in the United Kingdom by Netherwood Dalton & Co Ltd

JANE'S TRANSPORT PRESS

Cover illustrations

Front: Henson Airlines (PI)
Henson Airlines was founded in 1931 as a fixed-based operation.
Regular passenger flights began in 1962 under the name of
Hagerstown Commuter, linking Hagerstown with Washington DC.
In November 1967 Henson became the first airline to operate under
the Allegheny Commuter banner. It remained a member until 1983,
when it was taken over by Piedmont Airlines. Today the airline
operates to over 20 points in its network of scheduled passenger
services. These cover various destinations in Maryland, Virginia and
Pennsylvania, as well as to New York City and Washington DC. The
company operates a fleet of Beech 99, Shorts 330, DHC-7 and
DHC-8 aircraft. N906HA, a Series 101, is seen here just prior to
delivery, wearing its Canadian test registration. It was built in 1985,
and has construction number 009. (T Shwetz/De Havilland Canada)

Rear: Air Queensland (QN)
Air Queensland was formed on 21 June 1951 as Bush Pilots Airways.
Charter operations began that same year utilising a fleet of light
aircraft. Scheduled passenger services were added on 4 August
1957, and these now cover a network of over 35 destinations
throughout Queensland. Points stretch from Bamage in the north, to
Groote Eylandt in the west, and south to Brisbane. The company also
owns two holiday resorts, Lizard Island Lodge and Top of Australia
Wilderness Lodge. Air Queensland adopted its present name in
January 1982. A fleet of Cessna aircraft, DC-3, F-27, GAF Nomad,
ATR-42, and DHC-6 Twin Otter types are utilised. VH-TGH, a Series
300, is seen here wearing the company's striking livery.
(Udo and Birgit Schaefer Collection)

Above: Air Tanzania (TC)
Air Tanzania, the national airline of Tanzania, was formed in June
1977. Its aim was to operate services which had been suspended
following the collapse of East African Airways. The company's
network covers 21 domestic locations as well as international
destinations such as Kigati, Lusaka, Harare, Seychelles, Antanan-
arivo, Muscat, Djibouti, Dubai and Blantyre. Talks were underway in
mid-1987 for a link between Dar-es-Salaam and London, using
leased equipment. A fleet of Boeing 737, F-27, and DHC-6 aircraft is
maintained. 5H-MRB, a Series 300 Twin Otter, is seen here at
Shoreham, England, whilst on delivery to the airline. (R Finch)

Introduction

Announced in August 1964, the DHC-6 Twin Otter was designed as a STOL transport powered by two Pratt & Whitney PT6A-6 turboprop engines. Design work began in January 1964 with construction of an initial batch commencing in November of the same year. The first aircraft (CF-DHC-X) flew for the first time on 20 May 1965, with deliveries of the first production units beginning in July of that year. Designated as the Series 100, this model has sold over 115 examples. In 1968 the Series 200 had superseded its older brother. This type had an increased baggage hold and greater maximum landing weight. By the spring of 1969, DHC had created a larger model still, and therefore introduced the Series 300. This variant is a thoroughly developed utility aircraft with an outstanding record of versatility. It is the product of well over 20 years of design and development by de Havilland in fixed wing STOL and is a descendant of the successful Beaver, Otter, Caribou and Buffalo. As well as the passenger, cargo and combi DHC-6 types, there is also a maritime reconnaissance variant (-300MR), as well as a military transport and counter-insurgency type (-300M).

The DHC-7 Dash 7 quiet STOL airliner project was begun by de Havilland Canada in late-1972, following a worldwide market survey of short-haul transport requirements. Two pre-production aircraft were built. The first of these (C-GNBX-X) made its inaugural flight on 27 March 1975, with the second (C-GNCA-X) flying three months later. A third airframe was built for structural testing, whilst a fourth was used for fatigue testing. The first production Dash 7 (C-GQIW, c/n 3) flew on 30 May 1977. Rocky Mountain Airways (now Continental Express) was the first airline to put the aircraft into commercial service, doing so on 3 February 1978. Deliveries of the passenger Series 100 and the all-cargo Series 101 type have now passed the 100 mark. In 1985 a higher-weight, longer-range and greater fuel capacity model, the Series 150, was launched. It replaced its -100 brother on the production line in early-1986. Somewhat later than originally planned, the Dash 7 IR (ice reconnaissance) entered service with the Canadian Department of Environment in spring 1986. This one-off aircraft, registered C-GCFR, is a specially-equipped non-standard example of the DHC-7-150. It is intended for use in surveying sea ice and icebergs in the shipping and oil-drilling regions of the Labrador coast and Gulf of St Lawrence, when it supplements two Lockheed Electras already used for this purpose by the DoE's Atmospheric Environment Service. Non-standard features of the Dash 7 IR include a special dorsal observation cabin just aft of the flight deck, and a Canadian Astronautics SLAR 100 side-looking radar mounted in a fairing on the port side of the fuselage. This is used to locate ice in shipping lanes and drilling areas. Other mission equipment includes a laser profilometer to measure ice formation contours, photographic mapping equipment, and a data link between the aircraft and ships and drilling rigs in the patrol area.

The Dash 8 Series 100 is a quiet, fuel-efficient short-haul transport in the 30-40 seat category. The first of four flying prototypes (C-GDNK) made its first flight on 20 June 1983, followed by the second (C-GGMP) on 26 October, and the third in November. The fourth aircraft (the first with production PW120 engines) was flying by early-1984, followed in June by the first Dash 8 with production interior. Sized to accommodate 36-39 passengers, the Series 100 fits neatly between the company's 19-seat Twin Otter and 50-seat Dash 7. The first Dash 8 customer was NorOntair of Canada, whose first example of this type was delivered in October 1984. There are two basic examples of the Series 100; the Commuter, which is the standard passenger variant, and the Corporate, an 'executive' aircraft capable of carrying up to 19 people. There is also a Dash 8M which is a military version employed by the Canadian and American air forces. In 1985 a 'stretched' Series 300 model was announced. It is able to seat up to 56 passengers, carrying them up to 1482 km (921 miles). An inaugural flight was made in spring 1987, with deliveries scheduled to commence in the second half of 1988.

I would like to thank everyone who has helped in the compilation of the book, especially Ben Kooter of Vanwell Publishing, de Havilland Canada Public Relations Department, the DHC photographers, and finally, but by no means least, the Canadian High Commission, for its outstanding contributions. I would like to dedicate this book to Wendy Parkes, a very good friend.

TABLE OF COMPARISONS		
	DHC-6-100	**DHC-6-200**
Max accommodation	18	19
Wing span	19.81 m (65 ft 0 in)	19.81 m (65 ft 0 in)
Length	15.09 m (49 ft 6 in)	15.77 m (51 ft 9 in)
Height	5.66 m (18 ft 7 in)	5.66 m (18 ft 7 in)
Max t/o weight	4990 kg (11 000 lb)	5252 kg (11 579 lb)
Max cruis. speed	297 km/h (184 mph)	306 km/h (190 mph)
Maximum range	1480 km (920 miles)	1521 km (945 miles)
Service ceiling	7770 m (25 500 ft)	7400 m (24 300 ft)
	DHC-6-300	
Max accommodation	20	
Wing span	19.81 m (65 ft 0 in)	
Length	15.77 m (51 ft 9 in)	
Height	5.94 m (19 ft 6 in)	
Max t/o weight	5670 kg (12 500 lb)	
Max cruis. speed	338 km/h (210 mph)	
Maximum range	1297 km (806 miles)	
Service ceiling	8140 m (26 700 ft)	
	DHC-7-100	**DHC-7-150**
Max accommodation	54	54
Wing span	28.35 m (93 ft 0 in)	28.35 m (93 ft 0 in)
Length	24.54 m (80 ft 6 in)	24.54 m (80 ft 6 in)
Height	7.98 m (26 ft 2 in)	7.98 m (26 ft 2 in)
Max t/o weight	19 958 kg (44 000 lb)	21 319 kg (47 000 lb)
Max cruis. speed	428 km/h (266 mph)	428 km/h (266 mph)
Maximum range	2168 km (1347 miles)	4679 km (2907 miles)
Service ceiling	6400 m (21 000 ft)	6400 m (21 000 ft)
	DHC-8-100	**DHC-8-300**
Max accommodation	40	56
Wing span	25.89 m (85 ft 0 in)	27.43 m (90 ft 0 in)
Length	22.25 m (73 ft 0 in)	25.65 m (84 ft 3 in)
Height	7.49 m (24 ft 7 in)	7.49 m (24 ft 7 in)
Max t/o weight	15 649 kg (34 500 lb)	17 962 kg (39 600 lb)
Max cruis. speed	497 km/h (309 mph)	526 km/h (327 mph)
Maximum range	2130 km (1325 miles)	1482 km (921 miles)
Service ceiling	7620 m (25 000 ft)	7620 m (25 000 ft)

Previous page: **Brymon Airways** (BC)
On 15 June 1972 Brymon Aviation (as it was then called) made its first commercial flight from RAF St Mawgan (now also Newquay Civil Airport), using a nine-seater Britten-Norman BN2A Islander. The service was to the Isles of Scilly, just over 30 minutes away. From that inauspicious beginning at that tiny civil terminal, Brymon Airways grew into a scheduled operation. For the first year the company was based at Newquay, but as the route network grew Plymouth became the new base and remains the centre of operations. The first two routes granted were from Newquay and Plymouth to the Isles of Scilly and Jersey. Towards the end of 1972 the DHC-6 Twin Otter was introduced. 1973 saw the inauguration of the route to Guernsey, followed two years later by commencement of the Cork service. In addition, Exeter was introduced to the Isles of Scilly network. G-BFGP, a DHC-6 Series 300, is seen here during an approach to Plymouth. *(De Havilland Canada)*

Above: **ACES** (VX)
ACES, one of Colombia's three main domestic airlines, operates scheduled services between 38 destinations in the central, northern and western parts of the country. Jet flights cover trunk routes, whilst Twin Otter types run the numerous feeder lines. The privately-owned company was formed in 1971, with services beginning in February of the following year, using Saunders ST-27 aircraft. This type initially linked Manizales with Medellin and Bogota. Today, destinations that are served include Acandi, Guapi, Nuqui, Ocana, Turba, Urras and Yopal. Main operating hubs for regional operations are Bogota, Medellin and Cali. A fleet of Boeing 727, F-28, and DHC-6 Twin Otter types are utilised. HK-2381X, a Series 300, is seen here. It was delivered to ACES new in November 1979. *(N Chalcraft)*

Aeropostal (LV)

Aeropostal, a government-controlled Venezuelan airline, was established in 1930 as Compagnie Générale Aeropostale. Following nationalisation three years later, a name change was made to Linea Aeropostal Venezolana or Aeropostal for short. Operations were extended in mid-1957 when the airline TACA de Venezuela was absorbed. Today, the company operates both regional international and domestic scheduled passenger flights. A fleet of DC-9 and de Havilland Canada DHC-6 Twin Otters is utilised. DHC-6, YV-26C, is seen here at Miami International Airport. *(P Hornfeck)*

Air Atlantic (CP)

Air Atlantic commenced services in February 1986, with a base at St John's, Newfoundland. The first equipment used was the Dash 7 aircraft. The airline is partly owned by Canadian Airlines International, which has a 20 per cent interest in the company. Air Atlantic is a 'Canadian International Commuter', and feeds passengers from the Canadian airline through to destinations in the north of Canada. Points served include Stephenville and St John's (Newfoundland); Halifax, Yarmouth and Sydney (Nova Scotia); Moncton, Saint John and Fredericton (New Brunswick); and Charlottetown (Prince Edward Island). The company maintains its hubs at both St John's and Halifax. The fleet comprises both Dash 7 and DHC-8 types. An example of the latter can be seen here sporting a variation of the new and colourful Canadian livery.
(De Havilland Canada)

Air BC (ZX)

Air BC was formed on 1 November 1980 following one of the largest commuter airline mergers in aviation history. A total of seven companies were amalgamated into two operating divisions. Air BC combined Vancouver-based AirWest Airlines, Haida Airlines and West Air; also integrated were Campbell River-based Gulf Air Aviation and Island Air, as well as Nanaino-based Pacific Coastal Airlines. Trans Provincial Airlines, based at Prince Rupert, remained as a separate operation. Air BC maintains scheduled passenger commuter flights to various points in western British Columbia, as well as operating extensive charter passenger and cargo services. From its base at Vancouver International Airport, a flight hub is co-ordinated in conjunction with Canadian Airlines. A fleet of DHC-6 Twin Otter, DHC-7, Beaver, Otter, Goose and Cessna 185 aircraft is utilised. C-GJPI, a DHC-7, is seen here at Air BC's Vancouver headquarters.
(K Swartz)

Above: Air Creebec (YN)

Air Creebec was formed on 5 July 1982 to provide scheduled passenger services in western Quebec in association with Austin Airways. Passenger and cargo charter operations are also undertaken by the Val d'Or-based carrier. Points served include Eastmain, Paint Hills, Fort George and Great Whale. The company operates a single example each of the BAe 748, Cessna 402, DHC-6 Twin Otter Series 300, and the DC-3.

(Udo and Birgit Schaefer Collection)

Opposite: Air Guadeloupe (OG)

Air Guadeloupe maintains scheduled passenger services in the eastern Caribbean, connecting Guadeloupe, Antigua, Dominica, St Bartholemy, St Maarten and St Martin. Charter and contract operations are also undertaken to other destinations. The airline commenced services in March 1970. Current ownership control is held by Air France, which owns a 45 per cent share, while local public and government interests hold the remainder. As the national French flag carrier has such a large proportion of shares, Air Guadeloupe operates shuttle flights between Point-à-Pitre and Fort de France (Martinique) on its behalf. A fleet of ATR 42, Britten-Norman Islander, Fairchild F-27, and DHC-6 Twin Otter Series 300 aircraft is utilised. F-OGES can be seen here sporting the full Air Guadeloupe colour scheme. This aircraft was purchased from Miami Aviation Corp and was delivered in July 1970.

(Udo and Birgit Schaefer Collection)

Air Guinée (GI)

Air Guinée was established in October 1960 as a state corporation. This followed the signing of an agreement with the Soviet Union in March of that year for supply of aircraft and technical assistance. Operations began at the end of 1960 using Il-14 types over two domestic routes from Conakry to Boké and Kankan. Regional services followed shortly afterwards to Bamako in Mali, and some years later to Dakar in Senegal, Freetown (Sierra Leone) and Monrovia (Liberia). Additional destinations currently operated to include Abidjan, Lagos, Bissau, Faranah, Kissidaugou, Koundara, Labe, Macenta, N'Zerekore and Siguiri. A fleet of An-12, An-24, Boeing 707, 727, 737, Il-18, Yak-40, and DHC-7 aircraft is maintained. The last-named type, 3X-GCJ, a Series 102, can be seen here, just prior to delivery. *(T Shwetz/De Havilland Canada)*

Air Nova (CP)

Air Nova was formed in July 1986 by Air Canada, RWP Holdings Ltd, and Atlantis Corporation Ltd. With its headquarters in St John's, Newfoundland, and an operations base in Halifax, Nova Scotia, the company has a fleet of DHC-8 aircraft. The carrier has a partnership agreement with Air Canada which gives the latter a 49 per cent share in Air Nova. The national airline in return handles all ticketing, reservations, baggage transfers and connections at airports it serves. Although not fully operational in mid-1987, Air Nova expects to be handling over 300 000 passengers a year, with a load factor of 55 per cent. Destinations served include Goose Bay, Deer Lake, Halifax, Yarmouth, Moncton and Stephenville. At 805 km (500 miles) the company's Goose Bay-St John's service is one of the longest non-stop DHC-8 air routes.

(T Honeywood/De Havilland Canada)

Above: **Air West** (WI)

Air West was a commuter carrier that operated scheduled inter-city passenger services in Texas. The privately-owned airline made its inaugural flight on 25 December 1984, when it flew non-stop between West Houston/Lakeside Airport and Dallas Love Field. The company had its base in Houston and operated a DHC-7 fleet. Unfortunately, on 16 August 1985 due to various difficulties, Air West ceased operations. N210AW, which is seen here, was originally delivered to Wardair in June 1979. This Series 103 type was then sold to Air Wisconsin, then purchased by Air West in October 1984 as N791S. A registration change to N210AW took place on 1 December of that year. *(Udo and Birgit Schaefer Collection)*

Opposite: **Air Wisconsin** (ZW)

Air Wisconsin is a certified carrier that operates an extensive network of scheduled passenger, freight and mail services. The airline was formed in August 1965 and has grown into a large US regional airline. The company operates to points within the states of Illinois, Indiana, Michigan, Minnesota, Ohio and Wisconsin. A fleet of F-27, BAe 1-11, BAe 146 and DHC-7 types is operated. An example of the DHC-7 type is seen here during a flight to Kalamazoo. Air Wisconsin became the first airline to order the BAe 146-300. *(R Nunney/De Havilland Canada)*

Alaska Aeronautical Industries (YC)
Alaska Aeronautical Industries was founded in 1954 with a base at Anchorage. The company is one of the leading commuter airlines in the state and operates scheduled passenger flights from Anchorage to points on the Kenei Peninsula, Kodiak Island and in Denali National Park. AAI also offers charter flights and other non-scheduled services, including seasonal scheduled flights to Mt McKinley National Park. A fleet of DHC-6 and a sole Embraer 110 Bandeirante aircraft is maintained. N332MA, a Series 100 Twin Otter, is seen here at Anchorage.
(Udo and Birgit Schaefer Collection)

Alyemda — Democratic Yemen Airlines (DY)

Alyemda was formed on 11 March 1971 by Presidential decree as the national airline of the People's Democratic Republic of Yemen (formerly known as Aden). The company then acquired a fleet of DC-3 and DC-6 types and began operations from the capital. Today, scheduled passenger and cargo services are operated from Aden to Abu Dhabi, Addis Ababa, Al Ghayday, Beihan, Bombay, Damascus, Djibouti, Jeddah, Kuwait, Mogadishu, Qishn, Riyan, Sana'a, Seiyan, Sharjah and Socotra. A fleet of Boeing 707, 737, Tu-154B, and DHC-7 aircraft is operated. 70-ACK, a Series 103, is seen here at Aden in June 1981. Eleven months later the aircraft was the victim of an accident, when it was written off at the airport. The Dash 7 was originally ordered by Wardair in 1979 as C-GXVH, but it was subsequently cancelled and purchased by Alyemda. *(Udo and Birgit Schaefer Collection)*

15

Opposite: **Atlantic South East Airlines** (EV)

Atlantic South East Airlines, a rapidly-expanding regional commuter airline, was formed in June 1979. Through a merger in 1983, the carrier absorbed Southeastern Airlines. The company provides scheduled passenger services to over 20 points in Alabama, Florida, Georgia, Mississippi, North and South Carolina and Tennessee. Through a marketing agreement with Delta Air Lines, the company operates as part of the 'Delta Connection', co-ordinating flights with the company at Atlanta and Memphis air hubs. Although flights are designated via the DL ticketing code, ASA maintains its own identity. A fleet of DHC-7, Bandeirante, Brasilia, and Shorts 360s is utilised. N701GG, a DHC-7, is seen here at Atlanta in June 1982 sporting a previous livery. *(R Leader)*

Below: **Austin Airways** (UH)

Austin Airways, founded on 1 March 1934, is Canada's oldest continuously operated airline. It is also the country's largest third-level carrier. The company maintains an extensive network of scheduled services that links over 30 points in northern Ontario and Quebec. Daily international flights are operated between Thunder Bay (Ontario) and Minneapolis/St Paul (Minnesota, USA). Various charter flights are also carried out to other points in North America. Austin Airways' fleet consists of Beech 99, BAe 748, Cessna 402, DHC-3 Otter, DC-3 and DHC-6 Twin Otter aircraft. An example of the DHC-6, C-GNPS, a Series 300 variant, is seen here sporting the Austin colour-scheme. *(Udo and Birgit Schaefer Collection)*

Opposite: **Horizon Air** (QX)
Horizon Air, a Seattle-based regional carrier, was formed in May 1981. Following the company's acquisition of Air Oregon, in July 1982, the airline became one of the largest regional air carriers in the Pacific Northwest. Transwestern was acquired in January 1984, and Horizon introduced its first jet equipment in July of that year. The company operates a scheduled passenger network that serves over 22 points in Washington, Oregon, Idaho, Utah and California. The airline operates a fleet of Metro III, F-27, F-28 and DHC-8 aircraft. N811PH, an example of the DHC-8 type, is seen here just after engine start.
(T Shwetz/De Havilland Canada)

Above: **Brymon Airways** (BC)
In the autumn of 1981, Brymon introduced to the UK two examples of the DHC-7, as a long-term charter operation for Chevron Oil. They linked Aberdeen with Britain's most northerly airport, Unst. A third aircraft, based at Plymouth, was introduced the following year and is used for the London (Heathrow), Channel Islands, and Cork routes. In addition, a weekly Aberdeen-Plymouth scheduled service enables aircraft to be rotated for maintenance purposes. The 748 km (465 mile) non-stop service has a flying time of 2 hr 20 min. This has the honour of being the United Kingdom's longest route. Brymon was one of the first operators to fly from London's newly-opened City STOL-PORT. *(P Bish)*

Burma Airways (UB)

Burma Airways was formed in 1948 as Union of Burma Airways, a wholly government-owned airline. The company was established following the country's secession from the British Commonwealth. Domestic services were inaugurated between Rangoon and Mandalay using Douglas DC-3 aircraft. International routes were added in 1950 to Chittagong (in what was then East Pakistan) and Bangkok (Thailand). The company's present name was adopted in December 1972. Today a network of routes covers more than 20 domestic points, as well as international destinations to Calcutta, Kathmandu, Dhaka, Singapore and Bangkok. A fleet of F-27, F-28 and Aérospatiale Pumas is operated. Although no longer operated in the fleet, an example of a DHC-6 Twin Otter, XY-AEF, can be seen here at Shoreham Airport, UK. (R Finch)

CACC — General Administration of Civil Aviation of China (CA)

CAAC was established in 1962 as the General Administration of Civil Aviation of China. In addition to its airline operations, the carrier controls all civilian air transport activities in the People's Republic of China (including passenger, cargo, agricultural, airport, training and other specialised services). The government-controlled national airline maintains a comprehensive domestic scheduled flight network, as well as serving points within Europe, Asia, the Middle East, Africa, the USA and Australia. A 'mixed bag' fleet of aircraft is maintained, including examples of the de Havilland Canada DHC-6 Twin Otter. 512 is seen here at Shoreham, in the UK, whilst on delivery to Beijing. This photograph, taken in April 1978, shows the aircraft wearing its Canadian registration, C-GNOE. *(F Knight)*

Cape Smythe Air Service (6C)

Cape Smythe Air was formed in February 1975 as Fel Air, a Barrow-based charter operator. Scheduled services were started in mid-1979, and in November 1983 a second base was established at Kotzebue. Today the company operates scheduled commuter, charter and contract flights in northern Alaska, from its two hubs. Points served include Barter Island, Wainwright, Point Hope, Anaktuvick Pass, Deadhorse, Atgusuk, Point Lay and Nuiqsut. A fleet of Cessna Skywagon, Stationair 7 & 8, Navajo Chieftain, a Shorts Skyvan and four DHC-6 Twin Otters is utilised. N1022S, a Series 100, is seen here in the company's rather striking livery. *(Udo and Birgit Schaefer Collection)*

City Express (OU)

City Express is an expanding commuter air carrier that provides inter-city passenger services in Ontario and Quebec. Originally founded as Atonabee Airways, the company inaugurated services in January 1971 between Peterborough and Montreal. Following an ownership change in 1984, the City Express title was adopted. Major expansion also took place that year, with introduction of high-frequency daily services between Toronto Island Airport and Ottawa. The company operates a fleet of Saunders ST-27, DHC-7 and DHC-8 aircraft. A Dash 7, C-GGXS, is seen here at Toronto's Island Airport during a turnaround.
(Udo and Birgit Schaefer Collection)

Below: **Coral Air** (VY)

Coral Air, a privately-owned company, was formed on 15 February 1980 to provide services over a St Croix-San Juan route. In May of that year the airline absorbed Eastern Caribbean Airways. By 1981, however, Coral Air had become bankrupt, but re-emerged under new ownership in August of the following year. Further financial difficulties occurred, and in 1984 the airline ceased operations. At that time the company was serving St Croix and St Thomas in the Virgin Islands, and San Juan in Puerto Rico utilising a DHC-6 Twin Otter fleet of aircraft. N80701 is seen here whilst at St Croix. *(Udo and Birgit Schaefer Collection)*

Opposite: **De Havilland Canada DHC-6 Twin Otter**

The Twin Otter is a thoroughly developed utility aircraft with an outstanding record of versatility. It represents over 20 years of design and development by de Havilland in fixed-wing STOL. The aircraft is a descendant of the successful DHC-2 Beaver, DHC-3 Otter, DHC-4 Caribou and DHC-5 Buffalo. The basic type is the 'Commuter' version with 20 seats, in a three abreast configuration. The aircraft also comes in freighter, combi, float plane and ski variants. The DHC-6 has so far clocked up over 700 units in 80 countries. No other light transport in the 5670 kg (12 500 lb) class does such a variety of work in so many countries. On city strips or jungle clearings, the aircraft are at work around the clock every day of the year. A growing number of military forces and government agencies find a wide use for the Twin Otter, be it on wheels, skis, or floats. From the upper reaches of the Amazon to the snows of the Arctic and Antarctic, the aircraft wear their military markings with distinction. The DHC demonstration aircraft, C-GHDA, is seen here whilst on a visit to Shoreham Airport in the UK. *(R Finch)*

Above: **De Havilland Canada DHC-7 Dash 7**

In rugged Greenland Dash 7s equipped with cargo doors conduct mixed passenger and freight operations, replacing the less-productive rotary-wing equipment which once maintained the service. Along the fjords on the rugged western coast of Norway lie a number of small 800 m (2625 ft) airports. The Dash 7 is the only 50-seater aircraft capable of providing the isolated communities on this coast with regular, dependable air services. Through the inclement weather of Valdez, Alaska, the aircraft was the world's first ICAO Standard MLS operation with 6.2 degree approaches into a mountain-surrounded airport. In England, the Dash 7 is equally at home operating into London's (Heathrow), the world's busiest international airport. It can be found flying oil workers into a 610 m (2000 ft) airstrip in remote Shetland Isles, or utilising the London Docklands STOLPORT, for which it is ideally suited. Wherever the DHC-7 operates its acceptability makes it ideal for commuter services. The prototype Dash 7 is seen here whilst operating a worldwide demonstration tour. *(R Finch)*

Opposite: **De Havilland Canada DHC-8**

For over 50 years de Havilland has designed and manufactured aircraft that have anticipated the needs of operators and their passengers. The Dash 8, the newest DHC type, continues this long-established tradition. In service since late-1984, by mid-1987 it had sold over 100 units. The Series 100, with a passenger capacity of 40, is an ideal commuter aircraft, and operates for many airlines of such a nature. Its larger sister, the Series 300, flew in spring 1987. This aircraft seats up to 56 and initial deliveries will commence in the third quarter of 1988. The Dash 8 uses Pratt & Whitney PW120 and PW123 engines for the two types respectively. The Series 100 demonstrator aircraft, C-GGPJ, is seen here during its first flight. *(De Havilland Canada)*

Eagle Air (Arnaflug) (IS)

Eagle Air was formed in April 1976 as a private company to operate international charter flights. Operations commenced in June of that year from a base at Reykjavik. In 1979 the airline was authorised to operate scheduled passenger services from the Icelandic capital to 11 domestic points, as well as to Amsterdam, Dusseldorf and Zurich. A fleet of DC-8, Boeing 707, 737, and a sole DHC-6 Twin Otter aircraft is utilised. TF-VLE, a Series 100, was delivered to Eagle Air during December 1979. The aircraft is seen here at Reykjavik.
(Udo and Birgit Schaefer Collection)

Eastern Metro Express (EA)

Eastern Metro Express was formed in 1985 as a subsidiary to Metro Airlines. Its aim was to operate BAe Jetstream 31 and DHC-8 types on feeder services to Eastern Air Lines. With a hub at Houston Intercontinental Airport, the airline flies to points that include Albany, Knoxville, Columbus, Mont-gomery and Wilmington. Due to the close association between EME and Eastern, the company uses the latter's two letter designation 'EA'. DHC-8-101, N807MX, was delivered from de Havilland Canada during June 1986. The test registration C-GESR was utilised by this aircraft (construction number 41). *(Udo and Birgit Schaefer Collection)*

Emirates Air Services

EAS was formed in April 1976, with a base at Abu Dhabi. The company undertakes passenger and cargo charter flights, as well as contract services. These are operated within the United Arab Emirates and to regional Middle East points. Activities include executive flights, as well as support services for oil and gas production and exploration companies. A fleet of DHC-6 and DHC-7 aircraft is operated. A6-MRM, a Twin Otter, is seen here at Shoreham, whilst on delivery to the UAE. *(R Finch)*

Empire Airways (EM)

Empire Airways was formed in 1977 with a base at Hayden Lake, Idaho. The company provides daily scheduled commuter services between Boise and Coeur D'Alene, together with cargo and passenger charter and contract operations. A fleet of various Cessna aircraft, a DHC-6, a Piper Seneca II, and a Swearingen Metro II, is operated. Additional Twin Otters, as well as Beech 99 aircraft, are leased during the summer months to transport personnel to and from forest fires, under contract to the United States Forest Service. N331CC is seen here at Boise. *(Udo and Birgit Schaefer Collection)*

Below: ERA Aviation

ERA is a major company that operates fixed-wing aircraft and helicopters throughout Alaska. A rotary-wing service is also maintained along the coast of the Gulf of Mexico. Established in 1948, ERA is a division of the Houston-based Rowan Companies. Aviation operations are performed through a number of subsidiaries — ERA Helicopters Alaska Division, ERA Helicopters Gulf Coast Division, and Jet Alaska. The latter operates charter and contract flights with turboprop and executive jet equipment from bases at Anchorage and Deadhorse. The company operates a mixed fleet of helicopters as well as Convair 580, DHC-6 Twin Otter, Learjet, and a sole Dash 7. N27AP, a Series 103, is seen here at Anchorage.
(Udo and Birgit Schaefer Collection)

Opposite: Ethiopian Airlines (ET)

Ethiopian Airlines was formed in December 1945 as the national airline of Ethiopia. It was founded by Imperial proclamation to develop international services and to establish connections from the capital, Addis Ababa, to communities in isolated mountain regions where a very sparse surface transport system exists. On 8 April in the following year, five war-surplus C-47s were purchased and utilised over a local network and to the neighbouring countries of Yemen, Djibouti, Sudan, Kenya and Egypt. Today the airline also serves destinations in India, China, Africa and Europe. A fleet of Boeing 707, 720, 727, 767, DC-3, ATR-42, Cessna Skyhawk, DHC-5 Buffalo and DHC-6 Twin Otter Series 300 types is operated. ET-AIN, an example of the DHC-6 type, is seen here sporting its Canadian test registration.
(T Honeywood/De Havilland Canada)

Greenlandair (Grønlandsfly) (GL)
Greenlandair was established on 7 November 1960 by the Greenland County Administration, the Royal Greenland Trade Department, the Cryolite Mining Company Oeresund A/S and SAS. Its aim was to develop the vital communications needed between main centres on the island. The company holds sole concession to operate domestic scheduled services together with international and domestic charters. Operations began on 1 May 1962 with Convair Catalinas and a DHC-3 Otter, leased from the Canadian airline EPA. Greenlandair's present network includes the west coast communities of Nanortalik, Narssaq, Narssarssuaq, Julianehåb, Groennedal, Frederikshåb, amongst many others. The company also undertakes many varied activities, such as DEW-line (Distant Early Warning) supply flights for the USAF in Greenland and eastern Canada. Also included are ice patrols, search and rescue operations, geological surveys, and off-shore operations on the island and the North Sea. A fleet of Bell 212, S-61N and two DHC-7-103 types is utilised. The latter types can be seen here at Godthåb, with OY-CBT in the foreground and OY-CBU behind.
(De Havilland Canada)

Hawaiian Air (HA)

Hawaiian Air was established in 1929 as Inter Island Airways, and commenced scheduled flights between Honolulu and Hilo on 11 November of that year. Today the airline operates high-density scheduled services that connect all six Hawaiian islands, as well as providing international flights to American Samoa and Tonga. The carrier utilises a fleet of L-1011, MD-80, DC-9, and DHC-7 aircraft. An example of the latter, N929HA, a Series 102, is seen here at one of the island airports. This aircraft has been in service with Hawaiian Airlines since May 1981, when it was purchased new from de Havilland Canada.

(Udo and Birgit Schaefer Collection)

Opposite: **Loganair** (LC)

Loganair was formed in 1962 by Duncan Logan Contractors. In 1983 British Midland Airways acquired a majority controlling interest in the carrier, when it purchased shares from the Royal Bank of Scotland. The company operates extensive scheduled passenger services on the mainland of Scotland, and in the Shetland, Orkney and Hebrides islands. Points are also served within England, Northern Ireland and the Isle of Man. Loganair offers charter and contract services throughout the United Kingdom, and maintains the vital Scotland Air Ambulance Service. The carrier is unique as it operates the world's shortest scheduled commercial flight in the world, between Westray and Papa Westray in the Orkney Islands. This service has a flying time of only 2 min. A fleet of Shorts 360, F-27, DHC-6 Twin Otter, and Britten-Norman Islanders is operated. DHC-6, G-BGPC, is seen here at Luton Airport on 29 April 1984. *(A J Mercer)*

Below: **Maersk Air** (DM)

Maersk Air was formed in 1969 to operate IT and charter flights. Operations began in January of the following year using F-27s and a BAe 125. The carrier is a subsidiary of the enormous AP Møller shipping company. Today Maersk owns a large fleet of Super Puma, Bell 212, Boeing 737, DHC-7 and Fokker 50 aircraft. Scheduled services are operated within Denmark and the Faeroe Islands. The airline owns a 38 per cent share in Danair, and operates various services out of Copenhagen using DHC-7 and Boeing 737 aircraft. An international route is operated by Maersk between Billund and Southend (UK). DHC-7-102, OY-MBE, is seen here at Kastrup on 21 July 1982. *(A J Mercer)*

Malaysian Airline System (MH)

MAS, the government-owned national flag carrier of Malaysia, was formed on 3 April 1971 to be the Malaysian successor to Malaysia-Singapore Airlines (MSA). Operations of its predecessor began in 1947 under the banner Malaysia Airways. However, due to political reasons, a split in operations forced the countries to adopt their own national carriers. Malaysian Airline System operates to over 22 foreign destinations in Europe, Australia, the Middle East, South and East Asia, as well as the USA. A fleet of Boeing 737, 747, DC-10, A300, F-27 and DHC-6 Twin Otters is utilised. 9M-MDK is seen here at Shannon whilst on its delivery flight. *(W Cluitt)*

Manx Airlines (JE)

Manx Airlines was established in 1982 by British Midland Airways and the British Commonwealth Shipping Group, which own a 75 per cent and 25 per cent shareholding respectively. The company is an expanding British regional carrier that operates scheduled passenger flights from its base at Ronaldsway Airport, Isle of Man, to points in England, Scotland, the Channel Islands, Northern Ireland and Eire. Manx Airlines has operated a fleet of Shorts 360s and a single Vickers Viscount. The latter type was replaced in 1987 by an 'all-new' BAe ATP. Although no longer operated, de Havilland DHC-6 Twin Otter, G-BEJP *Skianyn Vannin* was leased from Loganair between the end of October 1983 and March 1985. The aircraft can be seen here at Glasgow Airport. *(P Bish)*

Opposite: **Merpati Nursantara Airlines** (MZ)

MNA is the nationalised domestic airline of Indonesia, formed in September 1962 by the Indonesian government. Its primary objective was to take over the network of internal services which had been developed by the KLM subsidiary De Kroondiuf. The following year the airline launched scheduled services. These now link Ujung Pachany, Surabaya, Jayapura, Jakarta and Biak over an extensive network within the numerous islands which make up the Republic of Indonesia. Over 100 points are served, including international destinations in Australia, (Perth) and Malaysia (Kuala Lumpur and Kuching). A fleet of BAe 748, F-27, C-212 Aviocar, Vickers Viscount and DHC-6 aircraft is maintained. PK-NUS, a Series 300 Twin Otter, is seen here.
(De Havilland Canada)

Above: **NewAir** (NC)

NewAir was originally established as an air taxi under the name New Haven Airways in 1962. The carrier launched commuter operations between New Haven and Islip in May 1978, with the current name being adopted in July 1980. The company operated flights in northeast USA, connecting points in Connecticut, Maryland, New Jersey, Washington DC and Pennsylvania. With effect from 20 February 1985 the airline was merged into Pilgrim Airlines, and now operates as Pilgrim/New-Air. At the time of the merger the company was operating a fleet of Bandeirante, Twin Otter and Shorts 360 aircraft.
(Udo and Birgit Schaefer Collection)

Opposite: **Newmans Air** (ZQ)

Newmans Air is an independent airline which was at one time owned by Newmans Group Ltd, the biggest ground transport concern in New Zealand. Scheduled passenger services were inaugurated on 12 February 1985 over a four-city network between Christchurch and Queenstown on South Island, and Auckland and Rotorua on North Island. In 1986 an agreement was reached with Ansett to form a company which would compete directly with Air New Zealand and its domestic services. Ansett-Newmans was formed and the company

now operates two DHC-8 aircraft. Although they are no longer in service, the airline initially operated two DHC-7 types. C-GFRP is seen here at Downsview, Ontario, prior to delivery and being placed on the New Zealand register. This aircraft is now in service with the British regional airline Eurocity Express. *(T Honeywood/De Havilland Canada)*

Above: **North Continent Airlines** (2N)

North Continent Airlines is a cargo airline that operates scheduled services in southern and central California, as well as to Arizona and New Mexico.

The carrier also performs charter and contract services. Founded in 1983, North Central operates two Hamilton Westwind IIIs and a DHC-6 Twin Otter. The former are converted from Beech E-185s, whilst the latter used to be a Series 100 but has been rebuilt as a -200 model. Points served are Albuquerque, Burbank, Fresno, Los Angeles, Phoenix, San Francisco, San Jose and Tucson. A base is maintained at Long Beach.
(Udo and Birgit Schaefer Collection)

NorOntair (NR)

NorOntair has been flying since 18 October 1971. The company, with a base at North Bay, maintains scheduled passenger services between 21 communities in the Canadian province of Ontario. The airline is owned by the local government and is run by the Ontario Northland Transportation Commission. Aircraft are operated and maintained under contract by three private air carriers — Bearskin Airlines, Air Dale and Austin Airways. NorOntair operates a fleet of DHC-6 Twin Otter and DHC-8 aircraft. The company became the first carrier to take delivery of the DHC-8, when it did so in 1984. C-GCJB, a DHC-8-101, is seen here sporting NorOntair's colourful livery.
(T Shwetz/De Havilland Canada)

Pelita Air Services (EP)

Pelita is the commercial flight division of Pertamina, Indonesia's national oil corporation. The company undertakes extensive support operations for its parent company, as well as providing charter, contract and leasing services. For nearly a decade, Pelita has been engaged in government 'trans-migration' flights. This is a vast operation that relocates many hundreds of thousands of Indonesians from heavily-populated Java to other islands in the Indonesian archipelago. Following a restructuring of Pertamina in 1978, Pelita's operations have been streamlined and the airline's financial structure tightened. With a base at Jakarta, the company operates a fleet of F-27, F-28, BAe 146, L-100-30 Hercules, DHC-7, Transall C-160P, Skyvan, C-212 Aviocar aircraft, as well as varied rotary-wing types. DHC-7, PK-PSX, is seen here at Shannon during 1983, whilst on delivery to Indonesia. *(W Cluitt)*

Ptarmigan Airways

Ptarmigan Airways was established in 1961, with a base at Yellowknife. The company operates local scheduled flights in the Great Slave Lake region of Canada's Northwest Territories. Passenger and cargo non-scheduled operations are also flown using wheel/ski and float equipment. Ptarmigan utilises a fleet of DHC-2 Turbo Beavers, Aztec, Cheyenne I, II and IIIs, as well as the 'old faithful' DHC-6 Twin Otter Series 300. C-FWAB, seen here with tow bar attatched, was delivered in 1980, having previously been with Wardair as CF-WAB. *(Udo and Birgit Schaefer Collection)*

Rio Airways (XO)

Rio Airways, one of America's largest commuter airlines, began commercial flight activities in 1967. The company's initial network covered southern Texas, operating from a base at Corpus Christi. In 1970 the carrier merged with Killeen-based Hood Airlines, and by 1973 had shifted operations to the northern part of the state, with a base at Dallas/Fort Worth. By 1977 Rio had yet again enlarged its fleet when it purchased Davis Airlines in November of that year. Today the airline operates an all Beech 1900 fleet. Although the carrier no longer utilises the DHC-7, N53RA can be seen here wearing the company's full livery.

(Udo and Birgit Schaefer Collection)

Royal Hawaiian Air Service (ZH)

Royal Hawaiian Air Service, the largest commuter carrier in the Hawaiian islands, maintains scheduled passenger flights to 11 destinations on five islands. The company was formed in April 1965 by Peacock-Woods Corporation, and began air taxi services from a base at Kona. The following year scheduled services were introduced. In 1969 the parent organisation sold the company to Royal Hawaiian Airways, a newly-formed subsidiary of Delaware-based Lumber Industries. Today the company serves Honolulu on Oahu; Kaunakakai and Kalaupapa on Molokai; Lanai City on Lanai; Hana, Kaanapali and Kahului on Maui; and Hilo, Kona, Kamuela and Upola on Hawaii. Royal Hawaiian operates a fleet of Cessna 402 and two DHC-6 Twin Otters. N202RH, a Series 100, is pictured here. *(Udo and Birgit Schaefer Collection)*

De Havilland Canada DHC-8 Series 300

The DHC-8 Series 300 prototype, C-GDNK, made a successful inaugural flight on 15 May 1987 from its base at Downsview, Ontario. The aircraft is an extended fuselage version of de Havilland Canada's 37/40-seat twin-engined DHC-8 Series 100. The type has already proved extremely popular with regional airlines, corporate customers and military users alike. The Dash 8 prototype is currently involved in an extensive flight test programme. This will lead to FAA and Transport Canada certification in the second half of 1988. Fitted with a straightened wing, stronger undercarriage and uprated engines, the Series 300 will carry between 50 and 56 passengers. *(De Havilland Canada)*

49

St Lucia Airways (SX)

St Lucia Airways is a privately-owned company which was founded in 1975. Its aim was to provide general charter services, including tourist flights, from St Lucia in the West Indies to Martinique and Barbados. A local passenger shuttle operation is flown within St Lucia, between Vigie Airport (near Castries in the northwest) and Hewanorra Airport (near Vieux Fort in the south). In 1984 an L-100-20 Hercules was acquired for cargo services between the island, Miami and Houston. However, world-wide charter cargo flights are now undertaken with the addition of jet aircraft. The current airline fleet consists of Boeing 707 freighter, Britten-Norman Islander, Hercules, and DHC-6 Twin Otter aircraft. N63119, a Series 300, was purchased by St Lucia Airways in 1985. This machine is no longer operated and has been replaced by J6-SLP, which is also a -300 model.

(Udo and Birgit Schaefer Collection)

Seair Alaska Airlines (KJ)

Seair, formerly Sea Airmotive, commenced operations in 1939, and was incorporated in 1951. Today the company is rapidly expanding and provides scheduled services to more than 60 points in Alaska, with route concentration in the southwest of the state. The airline also undertakes charter services to Canada, as well as air-taxi work within its home state. Seair has its main flight and route hubs at Anchorage, Bethel, St Mary's, Aniak and Kodiak. Main operations covered by daily or weekday services are Anchorage-Aniak-Bethel, Anchorage-Bethel-St Mary's and Anchorage-Iliamna. A fleet of various helicopters are operated, as well as DHC-2 Turbo Beaver and DHC-6 Twin Otter aircraft. *(Udo and Birgit Schaefer Collection)*

Below: **Southern Jersey Airways** (AL/6J)
Southern Jersey Airways was formed in 1970 with a base at Atlantic City, New Jersey. The company operates a scheduled passenger network under the Allegheny Commuter banner, of which it is a part. In 1977 the carrier acquired Atlantic City Airlines, thus enabling it to connect Wildwood/Cape May and Atlantic City. The airline today flies to points throughout New Jersey, New York, Pennsylvania and Washington DC. A fleet of DHC-6 and DHC-7 aircraft is utilised, all of which wear the Allegheny Commuter livery, but with the addition of Southern Jersey Airways on the fuselage. Twin Otter, N105AC, a Series 300, is seen here at its base during a turnaround.
(Udo and Birgit Schaefer Collection)

Opposite: **Swedair A/B** (JG)
Swedair was formed in 1975 following the merger of Svensk Flygtjanst AB Swedair and Crownair AB. The company operates scheduled passenger services that connect Örebro, Borland and Copen-hagen. The carrier is actively involved in charter traffic, target-towing services, air taxi and business aviation, as well as survey flights. Non-flying activities include aircraft maintenance and repair, airport operation and aircraft marketing. Swedair also undertakes scheduled passenger services on behalf of SAS and Linjeflyg. A fleet of Saab SF-340, Cessna 404, MU-2B, DC-3 and DHC-6 aircraft is utilised. SE-GXN, a Series 300 Twin Otter, is seen here having just taken off from Bromma on 23 July 1982. *(A J Mercer)*

Opposite: **Talair** (GV)
Talair was established on 5 May 1942 as Territorial Airlines. Charter services were immediately begun from Lae and Madang. In 1968 scheduled traffic rights were gained in the New Guinea Highlands region. A rapid route and aircraft fleet expansion followed, assisted by the acquisition of Sepik Air Charters in 1971, Macair Charters (and its Solomon Islands Airways subsidiary) in April 1975, and Panga Airways in October 1977. It was not until 1975 that the present name was adopted. The company serves many points throughout Papua New Guinea. Destinations include Afore, Bundi, Cape Verde, Daru, Embussa, Green River and Ihu as well as many others. A fleet of Cessna 402, Britten-Norman Islander, DHC-6, Bandeirante and DHC-8 aircraft is operated. The first of the latter type, P2-GVA seen here, was delivered in December 1986. The aircraft is used on scheduled services throughout the country, serving many remote jungle airstrips. *(C Bryant/De Havilland Canada)*

Below: **Turk Hava Yollari (Turkish Airlines)** (TK)
Turkish Airlines was formed in May 1933 by the Ministry of National Defence. The company was known until 1956 as Devlet Hava Yollari (Turkish State Airlines), when it became an independent corporation. Today the airline operates scheduled passenger and cargo services from Ankara, Istanbul, Antalya, Izmir and Adana, to 11 other domestic points. International destinations are serviced throughout Europe, the Middle East and Far East. Charter flights to East Germany are also undertaken. A fleet of DC-9, DC-10, Boeing 707, 727, and Airbus A310 aircraft is operated. Seen here is one of three Dash 7s that were previously operated by THY. TC-JCJ is pictured at Amsterdam's Schiphol Airport, on its way back to de Havilland Canada. The aircraft, devoid of any colour scheme, was subsequently leased to Maersk Air of Denmark. *(Udo and Birgit Schaefer Collection)*

Time Air (KI)

Time Air is a regional airline that operates scheduled and non-scheduled passenger and cargo services in Canada's three western provinces. The company was formed in 1966 as Lethbridge Air Services, adopting its present name three years later. Ownership of the airline is currently split between 12 Canadian citizens and corporations. With a base in Alberta, Time Air operates over such routes as Lethbridge-Calgary-Edmonton-Grande Prairie, and Edmonton-Peace River-Rainbow Lake. The airline operates a fleet of CV-580, CV-640, Shorts 360, Beech 99 and DHC-7 aircraft. One of the latter type seen here (C-GTAD, a Series 102) in a somewhat dramatic pose, was delivered to Time Air in May 1980. *(T Shwetz/De Havilland Canada)*

Tyrolean Airways (VO)

Tyrolean Airways of Austria was founded in 1958 as Aircraft Innsbruck. The company adopted its present name in 1980. The airline operates scheduled services linking Innsbruck with Vienna, Graz, Frankfurt and Zurich. Additionally, scheduled flights between Innsbruck and Dusseldorf are maintained during the winter months only. Exten-sive inclusive-tour flights cover the Mediterranean area, as well as ad-hoc charter and ambulance services. Tyrolean, headed by Managing Director Fritz Feitl, operates DHC-7 and DHC-8 aircraft. The latter, OE-HLR, can be seen here at Innsbruck. It should be noted, however, that this aircraft has since been re-registered OE-LLR.

(Udo and Birgit Schaefer Collection)

Upali Aviation

Upali Aviation was established as a division of the Upali Group, which has interests in chocolate, beverage, electronics, soap and newspaper businesses. It is also one of Sri Lanka's largest companies. The airline has now discontinued all scheduled, charter and flight support domestic passenger services within the country. At one stage, services connected Colombo with Anuradhapura, Batticoloa, Jaffra and Trincomalee. With a base maintained at the capital, Upali operated a DHC-6 Twin Otter Series 300, a Cessna Citation, a Cessna 206, and a Bell 206 Jet Ranger. 4R-UAA was delivered in August 1980, but was sold four years later to Skywest Airlines of Australia. The aircraft is seen here at Colombo. *(P Bish)*

West Coast Air Services

Yet another airline amalgamated into Air BC during 1980, West Coast Air Services used a fleet of DHC-6 Twin Otter Series 200s. The company operated scheduled flights throughout the Canadian West Coast. C-GKNR (construction number 186) made its first flight on 10 December 1968, before being delivered to Aeronaves Alimentadores a year later. In February 1978 it was sold to a John Woods as N54539, before going to West Coast Air Services in June of that year. The aircraft is seen here at Vancouver on 16 June 1980, sporting the WCAS livery. *(A J Mercer)*

Above: **Westair Commuter Airlines** (OE)
Westair was formed in 1972 as Stol Air. With a base at Chico, California, the company is a subsidiary of WestCom Holding. The airline has its main flight hub at San Francisco International Airport and serves points that include Sacramento, Santa Rosa, Eureka, Concord, Stockton and Monterey. Besides scheduled operations, the carrier maintains various cargo services. The current fleet consists of Cessna 402, Bandeirante, Shorts 360 and DHC-6 Twin Otter types. N53AN, a Series 200, has construction

number 135. The aircraft was delivered in 1982, but is no longer with West Air. *(K Swartz)*

Opposite: **Widerøe** (WF)
Widerøe, Norway's oldest airline, was formed in January 1934 by Norwegian aviator Viggo Widerøe. Regular passenger and mail services began that same year over an Oslo-Kristiansand-Stavanger-Haugesund route. The main part of the company's revenue was derived from charter and contract activities until 1968, when DHC-6

scheduled flights were inaugurated on local routes. Services now cover well over 35 STOL airports specifically constructed for commuter-type operations. With a base at Oslo, Widerøe links the picturesque Norwegian coastal region, from Kirkenes in the far north to Bergen in the south. Today the airline is owned by Fred Olsen, Braathens SAFE, SAS and over 600 shareholders. A fleet of Twin Otter and DHC-7 aircraft is maintained. LN-WFL, a DHC-7-102, is seen here en-route to Oslo. *(T Honeywood/De Havilland Canada)*

Winlink (St Lucia) (WZ)

Winlink is a St Lucia commuter airline that flies scheduled passenger services in the eastern Caribbean, on routes that extend north to south from Dominica to Grenada. The company was formed in 1982 to serve the Windward Islands from a base at St Lucia-Vigie Airport. The carrier operates a sole DHC-6 Twin Otter, J6-SLG, a Series 300. A shuttle service is flown within the island, linking Vigie and Hewanorra airports. Besides the scheduled operations, the carrier also undertakes regional charter flights. J6-SLH, construction number 156, was delivered to the airline in August 1983 but was sold 14 months later. *(N Chalcraft)*

Above: **Trans World Express** (TW)
Trans World Express is a subsidiary of Trans World Airlines. The company was formed to feed passengers from the parent company's main route system. A number of operators make up the airline. Resort Commuter, for example, operates DHC-6 Twin Otters in full Trans World Express livery. N921MA is seen here at Orange County Airport awaiting a passenger load for a flight to Santa Ana. Resort

Commuter provides frequent services throughout California, linking Los Angeles and Santa Ana, Palm Springs and Catalina Island. *(A Clancey)*

Overleaf: **Continental Express** (WP)
Continental Express is the commuter subsidiary of Continental Airlines. It operates feeder flights from hubs at Denver, Washington DC and Houston. Various airlines make up the carrier including

Presidential Airlines, Royal Airlines, Sierra Pacific and Rocky Mountain Airways. DHC-7, N27RM, is seen here at Denver Stapleton Airport. The latter subsidiary operates a fleet of this aircraft type as well as the DHC-6 Twin Otter. The carrier maintains a base at Denver, and operates frequent flights to the ski resorts of Aspen, Steamboat Springs, Alamosa and Pueblo (Colorado) amongst many others. *(A Clancey)*